Survival Guide to I.T.

Survival Guide to I.T.

Learn what goes on in the Information Technology field...

Pat Sullivan

Writers Club Press

San Jose New York Lincoln Shanghai

Survival Guide to I.T.
Learn what goes on in the Information Technology field...

Writers Club Press
an imprint of iUniverse, Inc.

For information address:
iUniverse, Inc.
5220 S. 16th St., Suite 200
Lincoln, NE 68512
www.iuniverse.com

ISBN: 0-595-22159-9

Printed in the United States of America

This book is dedicated to the men and women of the information technology field that have given their lives and personal time to this career.

CONTENTS

SURVIVAL GUIDE TO I.T.

**Learn what goes on in the Information Technology field…
And the managers that they deal with!**

So you're thinking of a career in Information Technology and you want to break into the field. This is the book for you! The book will help you understand just a little of what to expect.

We've all seen the TV advertisements about I.T. getting the short end of the stick. Guess what, it's not far from the truth. In fact you'll find out just how crazy it can get.

For the I.T. professionals reading this you will love to see the politics and pain you deal with brought out for everyone to see. You may even relate to many things in the book from your own experiences. You are in a brotherhood that is unlike any other in the professional world today. Here are some of the areas that we are going to touch on in this book.

- Types of Managers (For obvious reasons no real names <grin>)
- Ways they hire you. (Tips and tricks)
- Ways they fire you. (and perhaps why they fire you?)
- Budgets
- Why do they hate I.T.
- Politics
- Deals
- Benefits of an I.T. Career

Weather you're looking to get into the field or you have been in the I.T. field for years, you will enjoy this book as you learn or relive some of the secrets of this close-knit group of professionals

WHAT IS I.T.?

To tell the truth, no one really knows what I.T. is anymore. In the past the I.T. had been divided into 2 major areas. You had processing information and you had communication of information. After some time of these to separate and defined fields, they started to merge together. This was the start of what we now call Information Technology or I.T. as it is known. I.T. is in a world of it's own and does not share any relative similarities to such departments as Finance or even HR. There are so many different fascists of technology and titles with different definitions that no one can keep up with them. You have specialties in areas such as:

- Internet/e-Commerce Development

- Systems Administration

- Operations

• Security

• Consulting & Systems Integration

• Data/Database Administration
• GroupWare

• Information Systems Audit

• Networking/Telecommunications

• Technical Support

• Applications Development

• Systems Analysis

The above list just scratches the surface and from that list alone, I can name about 50 different additional sub titles. It's not as easy as you might think to pick a career path. In addition to that, the definition of each title may change from company to company, as well as industry to industry. By the time this book is published it will need to be updated again.

Which career path pays the best

If you're basing your career choice strictly on money you're already in trouble. Most people that start in I.T. come from other departments and want to jump right into the big money that they think is in this career field.

I'm going to let you in on a little secret, the Helpdesk or technical support department is where most people start. That gives you the best chance at getting the basics of I.T. down and allowing you to be able to touch almost every area of I.T. It also gives you the opportunity to choose a path. But let's be real clear with the average salary, depending on the region of the country you are in, can vary from around $27,000 to $35,000 to start. This may vary of course depending on the area. And as we have experienced all to well, the economy has a significant influence on salary ranges. It is just common sense and the rule of supply and demand in action. When the economy is going well and there are more jobs than qualified people to fill them salaries go up.

I have listed just a few salary ranges that were published by a major Consulting firm on the east coast at the beginning of 2001:

- CIO $122,250—$200,750

- CTO $ 98,750—$152,500

- Director of I.T. $ 92,250—$125,500

- Applications Manager $ 76,500—$109,250

- Project Manager $ 74,000—$ 96,500

- Network Architect $ 72,250—$102,000

- Network Engineer $ 62,500—$ 85,250

- Network Administrator $ 48,000—$ 69,250

- Webmaster $ 58,000—$ 82,500

I still have not answered the one question that I am asked on a daily basis, and the one that you are probably most interested in as well. What path pays the best? Well to be honest with you the Administration side of the department pays the best...most of the time that is.

In case you are not familiar with some of the abbreviations I used in the salary range table above let me explain them. CIO is the short name for the company's Chief Information Officer. CTO is the short name for the company's Chief Technical Officer. Many companies may have either a VP – Information Technology, or a Director of Information Technology. Titles may vary but the concepts are the same. Companies continue to pay these positions from $100,000 to well over $250,000 depending on the package and size of the company.

CHANGE TIME

There will come a time in your career when you will have to make a decision that all of us have had to make at one time or another. Do I go into the management side of things or do I stay technical? That question is not as simple as you might think.

Now understand that going into management does not mean that you are not technical any more, but let's face it as you progress up the corporate ladder your time will be increasingly devoted to meetings, paperwork, and other management activities. The other side of that coin is that you will spend less and less time doing the hands on technical things that probably lead you to consider a career in I.T. in the first place. If you rise high enough on the corporate ladder there will even come a time when there is No more Hands on!

Once you have made that choice you can clearly see a direction to go. If you desire to stay technical then there are still positions that pay very well once you become proficient and gain some experience.

One such position that comes to mind is that of the Oracle Administration and they can make into the 6-figure range. There are a few more that hit that range as well. But for the most part those positions are few and far between and require many years of experience.

Another issue that must be addressed here is that the economy can affect you in so many different ways. This is one of those unforeseen things that blindside you when you least expect it.

Keeping a job in the I.T. world is the top of the list these days. With the economy-hitting bottom, as far as jobs are concerned, it takes a lot to keep reminding yourself how great I.T. is, and that you really enjoy what you do.

Layoffs are not taking just the redundant people that are on staff but the cream of the crop as well. There are very talented I.T. professionals waiting tables hoping for an opening in any company. Looking back 12 months ago almost any I.T. professional could command a high salary with very nice fringe benefits.

But don't abandon all hope, those days will return and the I.T. career field will again flourish and prosper. Don't forget that working 50-60 hours a week and being on-call (24X7) is normal for I.T. people. If you add that together and then divide by your salary you'll find that you won't be making as much as you might think. Then let's don't forget to add the stress factor into the equation and then you have the real deal.

Far to often people desire more money than they are worth. People always think very highly of themselves and the desire of their heart is to get a salary increase. It is not the amount that they desire but the scorecard they keep. Many of them use the salary to keep track how well they have done in their careers.

On top of that they are often asked to fill out a skills assessment test to track their progress in the I.T. field. Or maybe they are asked to complete this assessment to see if they have any holes in their staff that must be filled. These are the statements that upper management uses when handing out the test. Sometimes they are being truthful in their search for the knowledge of skill sets on their teams. But it can also be used in a different way that is far from revealing the truth.

Once this test is given and the management receives it they have a powerful tool in their hands. If it came to review time for an employee they could judge the staff by their personal scores and award raises according to the ranking they have devised. This is not necessarily a bad thing, but you should be a where of that aspect of its use.

You should also understand that when it comes to letting people go they will use that very same skills assessment test to let go the weakest employee in a given field and keep the more knowledgeable employee to do two jobs instead of one.

It can also be used to build a case for an employee that they do not want to have on staff any more. Some people will work overtime to build a case to get rid of someone that they do not like. It always seemed funny to me that some managers spend more time working on getting employees, they dislike, out of the company when they could have spent the time training and mentoring them to an expectable level. In so doing they would have built a future commitment with that employees.

I have added a generic Skills Assessment Test for you to see. This test is often very simple and presented in a way that puts the employee at easy. A statement might be made that there is no wrong answer on this test. They could also say something that would lead you to believe that this would give you additional help in getting the training you need. Maybe they would even state that this could be how you get that additional training class you have been asking for.

Skills Inventory

Skill level on a scale of
0 to 5

Name:

0 = no experience	
	Department:
1 = personal only	Position:
2 = novice business	
3 = intermediate	
4 = advanced	
5 = expert	

Years of experience		Education and Certifications
	Programming Languages, development tools, Applications	

Years of experience		Beginning	Intermediate	Advance
0.00	Basic	0	0	0
0.00	C Language	0	0	0
0.00	C++	0	0	0
0.00	Visual C	0	0	0
0.00	Visual Basic	0	0	0
0.00	SQL	0	0	0
0.00	Java	0	0	0
0.00	Universe	0	0	0
0.00	Foxpro	0	0	0
0.00	MAS90	0	0	0
0.00	Oracle	0	0	0
0.00	Registrar	0	0	0
0.00	Seibel	0	0	0
0.00	SAP	0	0	0
0.00	CRM	0	0	0
0.00	LMS	0	0	0

	Operating Systems			
0.00	NT Terminal Server	0	0	0
0.00	OS/2 Warp	0	0	0
0.00	UNIX	0	0	0
0.00	Windows 2000	0	0	0
0.00	Professional	0	0	0
0.00	Windows 2000 Server	0	0	0
0.00	Windows 3.x	0	0	0
0.00	Windows 95	0	0	0
0.00	Windows 98	0	0	0
0.00	Windows NT 4	0	0	0
	Hardware			
0.00	Intel based PCs	0	0	0
0.00	Macintosh	0	0	0
0.00	RAID	0	0	0
0.00	Sun	0	0	0
0.00	Communications	0	0	0
0.00	Routers	0	0	0
0.00	Storage Technology	0	0	0
0.00	Servers	0	0	0
0.00	Optical/CD-ROM	0	0	0
	Computing Environments			
0.00	Internet Computing	0	0	0
0.00	Intranet Computing	0	0	0
0.00	e-commerce	0	0	0
0.00	Client-Server	0	0	0
0.00	Remote Access	0	0	0
0.00	Remote synchrnization	0	0	0
0.00	Outsourcing	0	0	0
	WAN Products/services			
0.00	VPN	0	0	0
0.00	DSL	0	0	0
0.00	Frame Relay	0	0	0
0.00	ISDN	0	0	0
0.00	T1/T3	0	0	0

	Data Management Abilities			
0.00	Data Access and User Administration	0	0	0
0.00	Database Backup and Recovery	0	0	0
0.00	Database Capacity Planning	0	0	0
0.00	Database Design	0	0	0
0.00	Database Implementation	0	0	0
0.00	Database Monitoring	0	0	0
0.00	Data Warehouse	0	0	0
0.00	Customer Data Bases	0	0	0
0.00	Database Technologies	0	0	0
0.00	Access	0	0	0
0.00	Crystal Reports	0	0	0
0.00	SQL Server	0	0	0
0.00	Foxpro	0	0	0
0.00	COGNOS	0	0	0
	Desktop Application Packages			
0.00	Microsoft Excel	0	0	0
0.00	Microsoft PowerPoint	0	0	0
0.00	Microsoft Project	0	0	0
0.00	Microsoft Word	0	0	0
0.00	Netscape Communicator/Navigator	0	0	0
0.00	VISIO	0	0	0
	Development Abilities			
0.00	Business Analysis	0	0	0
0.00	Feasibility Analysis	0	0	0
0.00	Systems/Process Flows	0	0	0
0.00	Design Specifications	0	0	0
0.00	User Interface Design (GUI)	0	0	0
0.00	System Analysis	0	0	0
0.00	Systems Integration	0	0	0
0.00	Consulting	0	0	0
0.00	WEB	0	0	0
0.00	e-commerce	0	0	0

	Media & Graphics Technologies		
0.00	Adobe Illustrator	0	0	0
0.00	Adobe PageMaker	0	0	0
0.00	CAD/CAM	0	0	0
0.00	Document Management	0	0	0
	Business / Industry Knowledge		
0.00	Manufacturing	0	0	0
0.00	Sales & Marketing	0	0	0
0.00	Training	0	0	0
0.00	Order entry	0	0	0
0.00	Procurement/ Supply chain/Inventory	0	0	0
0.00	Engineering	0	0	0
0.00	Human resources	0	0	0
0.00	Communications	0	0	0
0.00	Finance & Accounting	0	0	0
0.00	Mergers & Acquisitions - due diligence	0	0	0
0.00	Mergers & Acquisitions - Integration	0	0	0
0.00	Management reporting	0	0	0
0.00	International business (currency, reporting etc.)	0	0	0
0.00	International computing (double byte)	0	0	0
	Management Abilities		
0.00	Corporate Leadership & Direction Setting	0	0	0
0.00	Technology Strategic Planning	0	0	0
0.00	Data Center Management	0	0	0
0.00	Service Center Management	0	0	0

	Network Management Abilities			
0.00	Network Administration	0	0	0
0.00	Network Analysis	0	0	0
0.00	Network Architecture and Design	0	0	0
0.00	Network Capacity Planning	0	0	0
0.00	Network Design	0	0	0
0.00	Network Hardware & Equipment Planning	0	0	0
0.00	Network OS Installation and Upgrade	0	0	0
0.00	Network Performance Tuning	0	0	0
0.00	Network Protocols	0	0	0
0.00	Network Security	0	0	0
0.00	Disaster Recovery	0	0	0
	Network Protocols & Standards			
0.00	DHCP	0	0	0
0.00	DNS	0	0	0
0.00	Ethernet	0	0	0
0.00	Fast Ethernet	0	0	0
0.00	HTTP	0	0	0
0.00	IP (Internet Protocol)	0	0	0
0.00	NetBEUI	0	0	0
0.00	TCP/IP	0	0	0
0.00	H.323	0	0	0
0.00	ODBC	0	0	0
	Network Technologies			
0.00	3Com	0	0	0
0.00	LAN (Local Area Network)	0	0	0
0.00	Network Routing	0	0	0
0.00	PC Anywhere	0	0	0
0.00	Proxy Server	0	0	0
0.00	RAS (Remote access server)	0	0	0
0.00	Windows NT networking	0	0	0
0.00	WINS	0	0	0
0.00	Cisco	0	0	0
0.00	Lucent	0	0	0
0.00	Nortel	0	0	0
0.00		0	0	0

	Planning and Project Abilities			
0.00	Business Cost Benefit Analysis	0	0	0
0.00	Business Formal Presentations	0	0	0
0.00	Budgeting and controls	0	0	0
	Systems Administration Abilities			
0.00	Systems Help Desk Management	0	0	0
0.00	Systems Security and User Administration	0	0	0
0.00	Systems Security Policies and Procedures	0	0	0
0.00	Systems Software Installation & Upgrade	0	0	0
0.00	Application Software Installation & Upgrade	0	0	0
0.00	Systems Integration	0	0	0
0.00	Applications Service Providors	0	0	0
	Systems Management Utilities			
0.00	Veritas Backup Exec	0	0	0
	Technology Specialty			
0.00	CAD/ CAM/ CAE	0	0	0
0.00	ERP (Enterprise Resource Planning)	0	0	0
	Web, Application & Messaging Servers			
0.00	Exchange	0	0	0
0.00	Microsoft Outlook	0	0	0
0.00	SMTP (Simple Mail Transfer Protocol)	0	0	0

HOW TO START

Most people enter the I.T. field in one of two ways. There is the academic route and the On The Job (OJT) training approach. In the academic route a person would start by getting a college degree in one of the I.T. areas, then look for a job. In the OJT approach you would start with an entry-level job such as a position on a helpdesk, then through self-study and formal classes either get a degree of one or more of the industry standard certifications. No matter which of these approaches a person selects, they usually already have an interest in computers. They may have a little knowledge from working on PC's out of their home or at school for projects. They might have even had a job during the summer repairing PC's for a local company. That's a great start and has launched many careers today.

Lets take a look at the academic approach first. Once a person gets to college they are faced with several questions.

- What classes do I take?

- Which ones will help in the real world?

- Am I only going to learn from a classroom environment?

- Will it be enough to get that great paying job when I'm done?

Here is something that you should understand about colleges and a four year degree versus certifications and experience. To get that 4-year degree you will be required to take classes that don't have anything to do with your direction of choice.

Most colleges have degree paths that you must take to earn a degree. Now don't get me wrong, I have an MBA and I am working on a Doctorate of Management. But how many classes in that degree path do you expect to use in real life? Some classroom environments are purely theoretical in nature. Which a must when thinking of designing fields such as networks or database structures.

Some are designed to teach you how to troubleshoot problems in a system environment. This type of classes is not only needed to meet the schools degree requirements, but in many cases will prove to be invaluable for future reference.

Let's move forward a little to your long awaited graduation from college. You have your Bachelor's degree in hand and the world is your oyster. (Or at least that is what the recruiting materials promised four years ago when you first applied for admission. The truth is that you have spent an enormous amount of money to get that degree and need to start paying off that student loan, or at the very least you need to pay your parents back. You get out the help wanted section of the local newspaper or start searching through one of the many on-line recruiting sites, looking for that perfect job for you.

But wait a minute. What is this? You soon discover that most of the jobs require one or two years of actual job experience. You shake your head and say "Well I have a degree". The I.T. Director or the H.R. person feels your pain but they need both. A degree and experience is needed and is the way that most HR departments weed through the enormous amount of job applications they receive. It's almost a two-edge sword with you needing experience for the job but you can't get anyone to hire you to give you that needed experience. We will get into the hiring issues a little later on in the book, but suffice it to say that you are very frustrated to say the least.

On the other hand, if your goal is to start with a job first and then get a degree you might look at getting some certifications. What should you get to ensure a great job? There is no magic bullet here, and there is no one certification that will get you in the door for every job opportunity. Well you could get one of the following:

- MCSE (Deals mainly with system operations for computer systems running the Microsoft family of operating systems and other office products)

- A+ (Deals mainly with hardware and pc maintenance)

- Cisco (Deals mainly with networking)

- CCIE (Deals mainly with routing, switching and communications)

These are just some of the industry standard certifications you can get, but just getting these certification(s) will not guarantee that you can get a job after you have completed the certification process. Before you run out, sign up and take the courses that are required to prepare for the certification test, you should know that it may take you 6-12 months to complete the classes and take the test.

Now you're certified and you apply for a job to pay off that bill on your credit card you used in taking those classes. The certification does help and can show some experience toward that field. You may still have a problem if you do not have enough experience but in most cases the certification will get you in the door. The I.T. people are always looking for someone that can walk in the door and become productive as soon as possible.

A third issue to be considered is one that is not often talked about, and that is the practice of hiring from within. Most companies preach that hiring from within is the best practice, and most of us have heard that in our careers. Now let's consider the following two questions:

- How often do you feel that is practiced?

- Is it a fair position change regarding pay increase?

Let me give you a view from inside. Many I.T. professionals that are reading this book can tell you that it is not universally practiced, and in fact the companies that do hire and promote from within are the exception and not the rule.

One of the reasons for this is, many companies feel bringing in new blood helps stimulate the employees that are already here. Additionally people new to the organization often see ways to improve things that those who have been there a while can't. Also executive management teams occasionally feel that they can get a little better than they already have. Then they not only have you but also a person as good or better than you in the same department. The most ironic thing is that in most cases they pay more for a person coming from the outside. On top of that the person that wanted the job on the inside will, in most cases, train the newcomer.

With that said let's look at hiring from inside the company's employee pool. H.R. looks at the employees that have applied for the position and they know there is a person that would fit perfectly into that position. In fact the person may already be doing the job on an interim basis. You have read over the job description and the job posting. Especially the salary range for the position you are interested in. You are excited about the new position and the nice bump in salary that comes with it.

Then they offer the job to you and you find that it is no more than a 10% increase to you. Man, you are mad and you want to know why you are being offered the job at such a low salary. The answer is very simple; this is the standard policy for almost every company.

H.R. will read you a policy that in essence states an employee cannot receive a raise in pay greater than 10% unless they have written approval from an accounting authority. And then it still may not reach the range that was advertised. If you talk to any of your I.T. friends you will find that to be a very real truth. Welcome to the game people.

WAYS THEY HIRE

Many people do not understand that interviewing is an art. Just about everyone you talk to have his or her own tricks and tips to getting a good deal. In the following paragraphs I will touch on a few that I have found to be helpful, but this is in no way a complete list. And with experience and practice you will develop a style that works best for you.

I.T. people can often come across as arrogant or self-centered in their field. They can also be stubborn in their pursuit of the truth and what's right. Well guess what, in this career field it does have it's place. When you are interviewing for an I.T. position they want the information and they want confidence that you can handle yourself.

You are stepping into one of the most highly stressed jobs in the business world today. When you are asked a question, brag on yourself because no one else will. But be prepared to back up the bragging with facts and details. Tell them of the times you have saved the company money. Tell them about the times you kept a system from crashing. Let them know you are not afraid to stay until the job gets done, and have done so in the past when the situation demanded it.

Now to the question of money and getting the salary you want. Make sure you get the salary you want up front. Don't let them fool you into thinking that we will review your performance after say 90 days or 180 days and expect it to lead to a large increase then. Remember the 10% rule I stated earlier.

In fact never tell them what you want for a salary until they tell you what they are willing to pay. The first one that talks price loses the war. Never tell them what you're making currently; it's not any of their business. I know…I know, they always say they need to know. Take this as a typical scenario for that type of discussion.

Interviewer: Well what are you looking for in a salary range?

Interviewee: I really don't think that we need to focus on that as yet. I want to ensure we have a great fit for the job first.

Interviewer: So what are you currently making at your job?

Interviewee: I am currently making industry standard for my position.

Interviewer: I really need to know that we are in the ballpark with the salary.

Interviewee: OK, it seems you have a perceived range, what is the range you have in mind?

Interviewer: $$$,$$$ to $$$,$$$ is what this position has as a range.

Interviewee: Well there you go, I'm in that ballpark so we can continue on.

Note: You should say that even if they are off by a little. They might still have some money to play with and they don't want to tip their hand.

APPEARANCE

Please do not go to an interview wearing something that looks like you slept in it. Don't laugh; I have interviewed so many people that have done that I can't even count the times.

I have had people come in dressed all in black with black leather pants and overcoat. Not good either as it sends a message of a dark side. Wait until you have been hired to wear that on a bad day.

Bottom line on this issue is to wear something that is nice, conservative and speaks of professionalism. You can't go wrong if you meet those three things.

MANAGERS

There are so many types of managers in this career field that I can't possibly list every one of them here. I will however touch on a few of the most important and most frequent ones you will encounter. For those of you who are already in this profession you will not only agree and laugh but will also want very badly to send me your updated list. That could end up being a book in itself, we will call it "**Manager Nicknames and what they mean**".<grin> So let's get started with the manager types you can expect to see.

"Do You Right"

The "do you right" manager is one we all love and cherish until we find out how well they do you. Here is a wonderful scenario that will teach you just what I mean.

Let's say you have been called into the office of your new manager. Mr. Do You Right has told you that he is putting you into a very important position and that you will go places and become a very valuable member of the team by filling this slot. He tells you something like "This is a very important position and one in which you can make some valuable contributions to the company. However, I can't give you an increase in salary right now because of the budget. But as soon as we get things straightened out I will do you right."

Now you have been in that job for a few months and then you go to him for some help with the salary issue. You say some thing like "Mr. Do You Right, I have been in this job for a few months now. I have done everything you asked for and have performed above and beyond. I am now a great systems administrator and I would like the title and pay to go with that. Can you do me right here?"

Well, Mr. Do You Right says, "You know I can't do that because HR won't let me give you that much of a raise and without the certification in that field I can't change your title. But if you get that certification I will see. When you get the certification I will also put you in for a 10% raise during your yearly review." You are stunned

by this turn of events and say something to the effect of
"Didn't you know this was the policy when you put me
into this position?" His response is to reassure you with
"Don't worry I'll do you right when the time comes."

Welcome to the mildest of the managers that you will
meet. This type of manager means to do good but is
more focused on getting the job done than in taking
care of their people. Their attitude is one of let's get the
job done and then see if we can help the employee out.
We have all seen them.

"Two Face"

The Two Face manager is one that will not always be
visible but is often one you learn about much later, after
you have left the job. This type of manager is one that is
always listening to what you say and then turning it
around to benefit them selves. We have all seen this type
of manager but might not have recognized it at the time.

This manager is one that will avoid confrontation with
almost anyone. This makes them a very real threat as
they look to be your friend and are always there to help.
They are always telling you to trust them and that they
are there for you. Be weary of the wolf that puts on

sheep's clothing. This type of manager steals your ideas and makes them their own. They feel no guilt for their actions, and if caught they simple feel bad that they were caught.

This type of manager may even send you into meetings that you were never invited to. Just to do their dirty work for them. You may even find out from other managers that this wonderful manager trashed you when it looked like someone was needed as a scapegoat. It always amazes me that people like that can sleep at night, but they seem to do very well. They enjoy seeing turmoil and people in a hurry. They may even talk to you in confidence about others while talking to others about you.

Avoid being familiar with a manager like this. Be professional and polite but never reveal your inter most thoughts, they will use it against you. This is one of the worst managers to work for, but they are out there and by being very careful you can stay alive and win the game. Work is work and friends are friends, but the two should seldom cross the line.

"The Bandit"

I once had a manager that we called the bandit, for reasons of respect I will not go into why, but let me tell you about "The Bandit".

This person was neither a two faced manager nor a manager that promised to "do you right". This manager was evil from the start. They made sure you knew they were in charge and that they made the rules. It did not matter that the rules were ever changing and that you were expected to follow them, regardless.

This manager would yell a lot and point fingers at you when you had nothing to do with the situation. An example would be if a piece of hardware went down, as sometimes happens, they would blame you for not knowing that it was about to go. You would be told how incompetent you are and that you better get your act together in a hurry.

It was things like this that made this manager feel important and in a position of strength. Everyone that worked for this manager hated that person and wished a house would fall on them as it did on their sister, in the Wizard of Oz.

This type of manager would send you into a frenzy everyday, leaving you to go home with knots in your stomach. The sad thing about this is that upper management never sees this manager as being bad, but as someone who can motivate his or her people into action. They usually remain in their position for some time and torture as many people as possible.

I still talk to some of my friends in that company and they say that "The Bandit" is still working in that same position. This manager will probably never change or move once they have made their place because it gives them a since of power and comfort.

"Ghost"

We call this manager the "Ghost" because she never seems to be around when you need her. She swoops in with how she wants things done and then leaves with the same swiftness.

You are often left with a look of puzzlement on your face as you try to sift through the information you have just received. You then try to put together something that is close to what she wanted. It is often not exactly what she wanted but you know you need a first pass to get things straight. You give her the first pass and then she gives you a look of dismay. By this time she is wondering if you even listened to what she said, even if it was 40 hours of work and you were only given thirty minutes to get it done.

She never write things down for you and she looks at you as if to say, "can't you remember this simple instruction?" Well to tell the truth it is never simple nor is it clear, and you feel worse when they are through with you the second time around.

I have also seen managers like this that may even tell you word for word what they want. You then go away feeling that you have it this time. When you return with your final result it is wrong again. Not because you failed to follow their directions, but because the instructions you were given had no chance of getting the desired results. They never realize that you see them as a manager that has no clue. It is not you, but their lack of communication ability that is the problem.

A favorite ploy they use is one that we all must watch out for. They may call you and dictate an e-mail or message to you over the phone. They want you to send it so that it makes you look strong and in-charge What they are really doing is covering themselves and making it look as if the email or memo came from you. That way if there is a negative response, it looks as if it was your idea.

Think about it, you have just sent out an e-mail that you do not agree with. Your name is on the e-mail and you reputation and word are tied to that content. How are you going to defend something you may not agree with? This type of manager does this so that they can either say yes I had him/her write that e-mail. Only if the community likes the e-mail, or they can say that they never knew it was coming and blame you if everyone hates it. Great way to straddle the fence.

In the long run those types of managers never stay in one place long. After about two years they leave you with the mess they have created and move on. They feel that the company is a mess and that they can't seem to get it changed. Ride these types of managers out; you will win in the long run.

"The Firewalker"

This is the manager that we have all been looking for. Many claim this title but few fit the bill. This is the manager that you would walk through fire for, at her very request. Let's look at why this type of manager commands such respect.

This type of manager puts the needs and desires of her employees first and foremost. That does not mean that they are a pushover or easy. That just means that they are fair and just in their decisions. They never let outside manager's come down on their employees heads. They never yell at people in public. They go by the golden rule of chastising in private and praising in public.

This manager always seems to have time to listen to their employees even if they are overloaded and facing pressing deadlines of their own. They stop what they are doing and give the employee their full attention during the conversation.

You might be asking yourself why is this type of manager so rare these days. The answer is not one we might like to agree with but the truth never the less. The employees, in most cases, rob a manager of their abilities and skills. How can this be you ask. The answer is very simple and very true. You have employees that back stab their manager while the manager might be trying to cover or protect their people. You have the employees that hound and complain about everything and nothing at the same time. You have employees that never like anything that is being done and feel that they should have received better.

These types of employees drive their managers to a state of anger and stress. The manager begins to feel that their employees do not care and are not loyal to them. This is the worst feeling a manager can have and makes them lose sleep at night. If this behavior continues, that manager will start to take on the traits of the other managers we have listed previously.

Understand that it takes a combination of both manager and employee to make a good I.T. shop run. Teamwork is the key to a great place to work. Never let others break down your department's ability to communicate and gel as a team. United you will stand but let a malcontent in and your team will crumble unless that person is removed.

$
BUDGETS

The life blood of any I.T. department is it's budget. This is the money needed to pay salaries, upgrade current systems, and add new hardware, software and other projects. While an adequate budget is an essential element of any successful endeavor, it is even more so for the I.T. Department. It is also one of the most misunderstood business functions by employees.

Many managers feel that I.T. is just a deep dark hole for money to fall into. The User Community (The people within your company that actually use the computers on their desks, and other IT services, to perform their duties are often referred to as the 'User Community") frequently feel that I.T. can run with little effort and little or no money. The sales force, as well as most other departments, often feel that I.T. is a drain on the overall company budget. The comments you will hear are always the same. I.T. just wants the biggest and the best of everything. They just want that because it is the newest technology out and they like to play with new things. Nothing could be further from the truth.

The problem lies in the lack of knowledge of the User Community. They think they know what I.T. needs but the problem is they know just about enough to be dangerous.

If you have not heard them already, you will soon hear many of the most impossible comments imaginable coming from otherwise intelligent and competent people. Comments that make you mad or upset just because they are spoken out of ignorance. Let's list a few of them so that you can be better prepared for the on slot.

Comments:
- Why can't you just push a button and change this screen interface?

- What do you mean you need to plan for this software?

- Why can't you just load this software on my PC and make it work?

- Why should I go through you to see if the software is compatible with the network? My staff is the only ones going to use it on the network.

- I don't want to spend any time on this, just make it work.

- I don't want to type in any password I just want to turn it on and it comes up. But it better be secure.

- What do you mean the Internet is down, fix it.

- Why does it take so long to send this 50-MB file over the Internet? Can't you make it go the instant I push the send button?

The list could go on for pages but I will not go on with this, as you need to stop laughing and continue to read on. The budget allows you to provide the infrastructure, projects, and support that the User Community will need for the up coming year. They may not know they need it so you will have to be their proactive brain for them.

Always make sure you allow for budget cutting when designing your budget. Often companies cut budgets for up to 30% right off the top. They do this without understanding what they have just cut. It is a common practice to cut a budget then hand it back to the director and say here is your budget for the year. Well let's look at that for a second and see if you understand what that means to the department.

A cut of say up to 30% is major and can cripple an I.T. department in a hurry. Finance will often say, "look we know that you can find cost savings in such areas as telecommunication costs and network infrastructure. Maybe you could reduce head count and just work smarter not harder". Are you shaking your head with puzzlement yet? Sure you are and so does the director that was just handed that new and improved budget.

Just wait until middle of the year and you are over budget with no help in sight. You must go to an executive staff meeting and explain why you are over budget. I know you think that you can tell the truth about how you were handed the budget and how you let them know that it was not going to be able to be adhered to.

Wrong, you are going to show that the company has grown or that the business needs are increasing and the additional budget adjustments are needed to support such and outstanding growth. Other wise they will find someone who will tell them that.

After the budget has been approved you will face the final hurdle after the review and cut of your budget. And that is the delivery of that budget to your staff.

Invariably there will be things cut out of the budget, and in harder the economic times the budget cuts will be even more painful. It is important to remember that the projects that were cut are someone's pet project and or brainchild. These are highly skilled, professional people but they are not thinking about the cuts as strictly interims of the company's over all budget and finances. What they see is a rejection of their hard work and effort. And not surprisingly in many case this leads to hurt feelings, disappointment, and in some cases even anger. When this happens it is the I.T. Manager's job to help them see that their project being cut was just a business decision and not a personal attack. It is also important to remember that the ones making the budget decisions are often not I.T. people. They frequently have very little knowledge of the implications of their actions. In any event budget cuts are a fact of life and after you get the final budget you have to roll up your sleeves and do the best that you can with what you have.

The second issue is that during the year things always crop up and hit your budget that you never counted on. Late invoices from last year that were never accrued for, is a favorite one for all managers.

Then you have the maintenance fees that you forgot as well. On top of that you have the staff member that still does not understand that you must stay within your budget. They often want things that seem little to them but impact your budget, never-the-less. You can explain until you are blue in the face and they will never understand.

There are those who understand but still think it should be done and can't see why you don't find a way. The budget process is a never-ending fight with no end or rest, year after year.

I have inserted a sample expense worksheet that you may encounter. This worksheet has most of the line items you will see when working with an I.T. budget. The costs associated with this expense worksheet are fictitious but the line item nomenclatures are correct.

I have taken the first months review to look at forecasting the entire year. If you feel that the first month is a good representation of the costs associated with the rest of the year you can forecast for the next quarter. This type of thinking maybe a little shaky as the first of the year is not a good indicator for everything. You still have carry over from the previous year that will make this a difficult task at best.

As you will see it does not speak to the capital budget, which is an entirely different budget and line item section itself. We will address this in a later book as it is an entire chapter by itself.

	Budget	Jan.01 Actuals	Jan Budget	Variance	Balance forecasted
Salaries	$ 3,370,556.00	$ 96,679.00	$ 280,879.67	$ 184,200.67	$ 3,273,877.00
Other Wages			$ -	$ -	$ -
Bonuses	$ 374,501.00	$ 18,369.00	$ 31,208.42	$ 12,839.42	$ 356,132.00
Fringe Benefits	$ 417,408.00	$ 15,469.00	$ 34,784.00	$ 19,315.00	$ 401,939.00
Social Taxes	$ 364,505.00	$ 9,227.00	$ 30,375.42	$ 21,148.42	$ 355,278.00
Life/Accident Insurance			$ -	$ -	$ -
Auto Operating Lease			$ -	$ -	$ -
Subscriptions & Professional Fees	$ 4,000.00		$ 333.33	$ 333.33	$ 4,000.00
Other Fees	$ 3,000.00		$ 250.00	$ 250.00	$ 3,000.00
Training & Seminars	$ 60,000.00	$ 200.00	$ 5,000.00	$ 4,800.00	$ 59,800.00
Relocation Expense			$ -	$ -	$ -
Travel - Air	$ 60,000.00	$ 1,141.00	$ 5,000.00	$ 3,859.00	$ 58,859.00
Travel - Lodging	$ 60,000.00	$ 2,232.00	$ 5,000.00	$ 2,768.00	$ 57,768.00
Travel - Meals & Entertainment	$ 13,000.00	$ 539.00	$ 1,083.33	$ 544.33	$ 12,461.00
Travel - Business Meetings	$ 13,000.00	$ 477.00	$ 1,083.33	$ 606.33	$ 12,523.00
Travel - Rental Car	$ 3,000.00	$ 102.00	$ 250.00	$ 148.00	$ 2,898.00
Travel - Personnel Mileage	$ 3,000.00		$ 250.00	$ 250.00	$ 3,000.00
Travel - Other Reimbursements			$ -	$ -	$ -
Other Employee Reimbursements			$ -	$ -	$ -
Office Supplies	$ 8,500.00	$ 173.00	$ 708.33	$ 535.33	$ 8,327.00
Equipment & Furniture Expense	$ 3,000.00	$ 6,906.00	$ 250.00	$ (6,656.00)	$ (3,906.00)
Software Purchase			$ -	$ -	$ -
Contract Labor 180K& Consultant Svc 140.2K	$ 330,000.00	$ 47,512.00	$ 27,500.00	$ (20,012.00)	$ 282,488.00
Freight	$ 6,000.00		$ 500.00	$ 500.00	$ 6,000.00
Rent Expense	$ 66,000.00		$ 5,500.00	$ 5,500.00	$ 66,000.00
Machinery & Equip. - Repairs & Maintenance	$ 60,000.00	$ 3,442.00	$ 5,000.00	$ 1,558.00	$ 56,558.00
Equipment Rental	$ 6,000.00	$ 967.00	$ 500.00	$ (467.00)	$ 5,033.00
Food Services	$ 6,000.00	$ 128.00	$ 500.00	$ 372.00	$ 5,872.00
Telecom. Usage Expense	$ 440,000.00	$ 54,639.00	$ 36,666.67	$ (17,972.33)	$ 385,361.00
Telecom. Equipment Rental	$ 60,000.00	$ 17,091.00	$ 5,000.00	$ (12,091.00)	$ 42,909.00
Leased Line Expense	$ 175,000.00		$ 14,583.33	$ 14,583.33	$ 175,000.00
Other Direct Telecom.			$ -	$ -	$ -
Hardware Maintenance	$ 6,000.00	$ 2,263.00	$ 500.00	$ (1,763.00)	$ 3,737.00
Software Maintenance	$ 60,000.00	$ 16,708.00	$ 5,000.00	$ (11,708.00)	$ 43,292.00
Software Licensing & Royalties	$ 240,000.00		$ 20,000.00	$ 20,000.00	$ 240,000.00
Miscellaneous Expense	$ 20,000.00	$ 15,053.00	$ 1,666.67	$ (13,386.33)	$ 4,947.00
Recruiting	$ 72,000.00		$ 6,000.00	$ 6,000.00	$ 72,000.00
Payroll Outsourcing			$ -	$ -	$ -
Outside Services		$ (2,125.00)	$ -	$ 2,125.00	$ 2,125.00
Other Professional			$ -	$ -	$ -
Depreciation - Computer Hardware	$ 155,000.00	$ 10,468.00	$ 12,916.67	$ 2,448.67	$ 144,532.00
Depreciation - Computer Software	$ 130,000.00	$ 50,650.00	$ 10,833.33	$ (39,816.67)	$ 79,350.00
Depreciation - Furniture	$ 130,000.00	$ 315.00	$ 10,833.33	$ 10,518.33	$ 129,685.00
Depreciation - Machinery & Equip	$ 99,000.00	$ 7,133.00	$ 8,250.00	$ 1,117.00	$ 91,867.00
Depreciation - Lease Hold Improvement		$ 1,061.00	$ -	$ (1,061.00)	$ (1,061.00)
Capital Equipment Write - off			$ -	$ -	$ -
Services from (internal)	$ 500.00		$ 41.67	$ 41.67	$ 500.00
TOTALS	$ 6,818,970.00	$ 376,819.00	$ 568,247.50	$ 191,428.50	$ 6,442,151.00

WHY DO THEY HATE I.T.

We have touched on many issues already for the reason I.T. professionals are hated or despised. In most cases it is not that they hate I.T. professionals but that they do not understand the field.

They don't like the fact that we need money to operate and in their eyes you do not produce revenue. As we all know they would not be able to do their jobs without us but they would never admit to that. You can never meet the deadlines that they want, and you always make them accountable for their changes and requests. You put them in a position of accountability and that is never good for them.

I.T. is often blamed for things because it deflects from their lack of ability to get their own jobs done. Let me give you a few examples that will illustrate what I am talking about.

A salesperson sends in a laptop to be repaired. The day that it arrives the support staff repairs and reloads it and has it out the very next day. The user complains that they could not get their job done because of I.T. The claim to the upper management was that it took a week to get their laptop back and that hindered their abilities to perform their job.

This was a true to life situation and now let me tell you the whole story. The laptop was sent in and arrived on the day of a major snowstorm. The company that was to deliver the package could not for two days, which turned out to be Thursday and Friday. They delivered it on the next Monday and the I.T. support staff had it in the mail on Tuesday. The user was without the laptop for a week.

But this is still not the end of the story as the real truth continues to come out here. The laptop needed to be repaired due to the fact that the user and purposely uninstalled the virus protection software on the machine. This caused a major virus to corrupt the laptop thus making it unusable. Had the user not deleted needed software in the first place a trip to support would not have been needed. The remote salesperson not only caused their own problem but also when it came right down to it they could have still visited people and made calls using paper, pencil and a phone. But the upper management never heard the entire story nor did they care to.

I.T. departments are an easy target for people that need the attention turned away from them. They can use excuses for everything such as my e-mail will not let me send a note. By the way I actually saw a case where a user used this excuse, and without thinking used their own computer to send the e-mail.<grin>

They may even use the "My in box is to full and I am losing sales because you will not increase the size of the box." Excuse. Then when you investigate you discover that the user had over 1800 e-mails in their inbox plus another 50 GB in other areas of their folders. They never liked the idea of deleting anything or even saving it to their hard drive. Don't even think about telling them about archiving. But it is always I.T.'s fault.

Now don't get me wrong, I.T makes their own mistakes and configuration problems as all humans can. There are also hardware and software failures, as well as issues that are outside their control. It all adds up to this being a very hard career field to keep your sanity in. They don't really hate you but you are the best scapegoat a company could have.

Deals

This section of the book might not be anything like what you think it is at all. It's not about the shady deals that go on to better one's position or income. This section is about the deals that you can sometimes make with vendors, suppliers, and other department managers to get projects completed or software purchased. As we talked about during our budget conversation, sometimes you don't get the all of the funding that you requested in your budget. But there are ways that you can get around some of the short falls you encounter.

Many times you can get a better price from a vendor than you had previously budgeted for. Also vendors will often have promotional deals going, such as two for the price of one, or buy one get another one at a good discount. Sometimes you can combine your needs with those of another department and get a better price per unit because of the increased volume. Use these situations to your advantage along with making great relationships with vendors who are looking for future sales. Showing the company a, "money savings deal", like this is something they can understand and approve with a little promoting.

We once got a deal by purchasing a server for our accounting department at regular price, which was in the project's budget; we also received a second one for a single dollar. This gave us the opportunity to complete a project that was not in the final budget. Remember there is more than one way to skin a cat.

POLITICS

To say that you can come in, do your job, and go home everyday without having to play the political game is a dream that will never come true. Politics is the inevitable consequence of people interacting with each other, whether it is at work or at play. At least in this one area the I.T. field is no different than any other. As an I.T. Manager you need all of the skill, at this game, that you can muster as you try to meet the needs of your User Community and do it within the budget the company's Senior Managers give you. And while you are balancing those two competing groups, you will need to find a way to take care of your employees and meet their needs as well.

As a manager there have been times when I have had to tell my employees to do things I knew was not the best way, and in some cases totally ineffective. And as you can well imagine I heard about it from my people. They were right to complain about doing something that we all knew was not the best technical solution or the best use of our resources. But some times you just have to do what you are told, get the job done, and hope for the best. There are a couple of reasons why you might have to do this.

The first of the two reasons is that after you have given your boss all of the information she decides on a course of action that you disagree with. You gave her all of the information and presented your case as powerfully as you can. In the end she is your boss and after she makes her decision you owe it to her (and to your own integrality) to do your best to make things work.

The second reason is that you have made one of those political deals that must be done sometimes. You agree to do something one way if they agree to use their budget to purchase something for you or sign off on another position. It might even be to allow you to move to better facilities to enhance your employees environment and work area.

Politics is not always a dirty word, but it is always interesting. Sometimes it works to your advantage, and sometimes you just have to bite your lip and look at the results it will bring. Let me close this section with a final thought. Whether you are greasing the wheels to get what you need, or simply doing what you have to; make sure that you always have a backup plan and a restore function in place.

WAYS THEY LET YOU GO

In this day and age this section could be filled with many stories of current events. The massive layoffs that are occurring have been horrific and terrible for everyone. This is especially true for those of you who have been laid off recently.

Being laid off can come in many different forms and is sometimes linked to the types of managers we have discussed in earlier sections. The results are always the same, sadness and confusion, and in most cases a shock to the employees being let go.

Some managers like making you wait until the end of the day to tell you that you are being let go, oh and by the way here is two weeks severance pay. Some like to do that on Fridays so that you have all weekend to worry and get over it. Some like to do it on a Monday because that gives you extra time to look for a new job. It just depends on the manager, as I've stated before.

Then you have the massive layoffs that have been happening in our current economical state. Some have come to work on a Monday to find that they had a note on the door to report to HR. While others have been locked out of the company's secure building. And still others were just locked out of their section.

How would you like to show up to work on a Monday to find that you seem to have difficulty getting into your office? Just to find out later that it's not your office anymore.

In normal times the manager and HR have already got their game plan in place before you are ever talked to. If it is a disciplinary action, that is a totally different conversation and one we will not touch on in this book.

The normal action of reducing head count based on budget or economical discussions is the focus here. Some managers are very reluctant to let an employee go and find it very difficult. On the other hand there may be cases where the manager can be down right happy about getting rid of someone. Downsizing is a good opportunity for them to get rid of their trouble employees.

But all is not lost if you are quick to ask for things and not show them that you are shocked at all. Let them think you always have a backup and plan for just such an occasion. Ask for extra severance if possible right off the bat. If the company policy is stated so that you know in advance that cannot happen, then use another tactic.

Let them know that you know they are giving you all the money they are allowed by policy but that it would help you enormously if they could give you free training or pay 100% of the medical for awhile during this transition. Make sure you know all the company perks that are out there even the ones that only the VP's have. They may be given to you if you can catch them off guard with your professionalism. Ask for the world, there is no shame at this point in the game of thinking of yourself and your family.

BENEFITS OF AN I.T. CAREER

We have talked about some of the things that you may face in this career field. It is by no means complete but it should get you thinking before you change your current career field into this one.

You have found out the pay may not be what you think which is always a good thing to know up front. You know some tricks to getting to the interview and what education track to take. You know about managers and the types of managers you may face. The money and the tight budgets are always there. You have even heard how other departments think about I.T. You might have even been one of those users yourself in the past.

But let me tell you some of the benefits to being in this career field that you need to know. We have talked about the bad now let's get the good out. Down the road you do have a chance to make some good money at a very young age. This in itself can be a plus while you use the extra income to increase you knowledge and life style. There are often bonuses tied to project goals and to meeting your budget requirements, this is always a nice perk to have.

The technology is always around you and you feel connected to the future. Knowing this is the direction of all companies, even if they don't know or understand that concept, makes you feel a little ahead of the rest of the world.

The I.T. career field is a relatively small group of people across America. It looks like it's not but you will start making friends everywhere. Distance no longer is a barrier to keeping in touch with newfound friends. You need these people as there is no one that knows it all and the more contacts you make the more resources you have to draw on.

The grapevine is great in I.T. as you find out how other companies are overcoming challenges. You don't have to re-invent the wheel, you can modify someone else's solution to fit your needs. You can even find out if there is an opening that you do not know about.

FINAL THOUGHTS

We use information technology to increase productivity, data distribution, revenue, and to create knowledge. There are many other reasons but these are some of the main thoughts.

A business in this day and age typically spend, on the average 4.5% to 5% of their revenue to maintain an I.T. department. Therefore, the steering committees of companies today must consider their I.T. budgets during their decision-making processes. Many companies' today face major business problems that require an I.T. solution. For instance, the increasing need for accounting and data storage resolutions.

One final thought I run into all the time is, when to say "I don't know". There is no I.T. professional in the world that can say; "they know everything". That would be an overstatement, besides no one would believe it anyway.

I surround myself with very smart and talented people because I don't need know everything; I just need to know who does.

ABOUT THE AUTHOR

Pat Sullivan has over 24 years in the electronics industry. He teaches for University of Phoenix covering such areas as Project Management and courses dealing with Project Implementations. He is the Director of I.T. for a major computer training company. He has an MBA in Technology Management from University of Phoenix and a Business undergraduate degree.

Glossary

Administrator:
A person responsible for managing the local area network (LAN). Other duties of this position can be configuring the network, maintaining the network's shared resources and security, assigning passwords and privileges and helps users.

Allocation of Resources:
A reason given which allows management to not do what they should do to grow their company.

Application:
A software program that carries out a useful task. Word processors, spreadsheets and communication packages are just a few.

Bandwidth:
The capacity to move information.

E-commerce:
Electronic Commerce or EC—is the buying and selling of goods and services on the Internet, especially the World Wide Web. In practice, this term and a newer term, e-business, are often used interchangeably.

Enterprise:
Enterprise is an organization that uses computers. A word was needed that would encompass corporations, small businesses, non-profit institutions, government bodies, and possibly other kinds of organizations. The term is applied much more often to larger organizations than smaller ones.

Firewall:
A combination of software and hardware which limits the exposure of a system to an attack from outside. Such as a LAN connection to the internet.

Groupware:
Software which runs on a local area network and allows users on the network to work together on a joint project.

LAN:
Local area network – is a geographically localized network consisting of both hardware and software. In most cases in the same building or office.

Information Technology:
A name for data processing, which became management information systems (MIS), which became information technology, but in the long run its all the same.

Network:
In information technology, a network is a series of points or node interconnected by communication paths. Networks can interconnect with other networks and contain subnet works.

The most common topology or general configurations of networks include the bus, star, and token ring topologies. Networks can also be characterized in terms of spatial distance as local area networks (LAN), metropolitan area networks (MAN), and wide area networks (WAN).

Online:
Online is the condition of being connected to a network of computers or other devices. The term is frequently used to describe someone who is currently connected to the Internet.

Systems management:
Systems management is the management of the information technology systems in an enterprise. This includes gathering requirements, purchasing equipment and software, distributing it to where it is to be used, configuring it, maintaining it with enhancement and service updates, setting up problem-handling processes, and determining whether objectives are being met. Systems management is usually under the overall responsibility of a Chief Information Officer (CIO). The department that performs systems management is sometimes known as management information systems (MIS) or simply information systems (IS).

Telephony:
Telephony is the technology associated with the electronic transmission of voice, fax, or other information between distant parties using systems historically associated with the telephone, a handheld device containing both a speaker or transmitter and a receiver.

TCP/IP:
TCP/IP (Transmission Control Protocol/Internet Protocol) is the basic communication language or protocol of the Internet.

0-595-22159-9

www.ingramcontent.com/pod-product-compliance
Lightning Source LLC
Chambersburg PA
CBHW051255050326
40689CB00007B/1200